How to use this book

INSTRUCTION
What your child needs to do for the activity.

TITLE
The page title describes the skill your child will learn in these pages.

FUN ILLUSTRATIONS
Specially drawn illustrations which are fun, interesting and drawn at the right pedagogical level for your child.

COLOURFUL BORDERS
The page borders make each page as attractive as possible to stimulate your child.

EXAMPLE
The first one is done for you so you can show your child exactly what to do.

LOTS OF PRACTICE
Two pages where your child can practise and repeat the same skill to master it.

STICKERS
Place a sticker on each page as your child finishes.

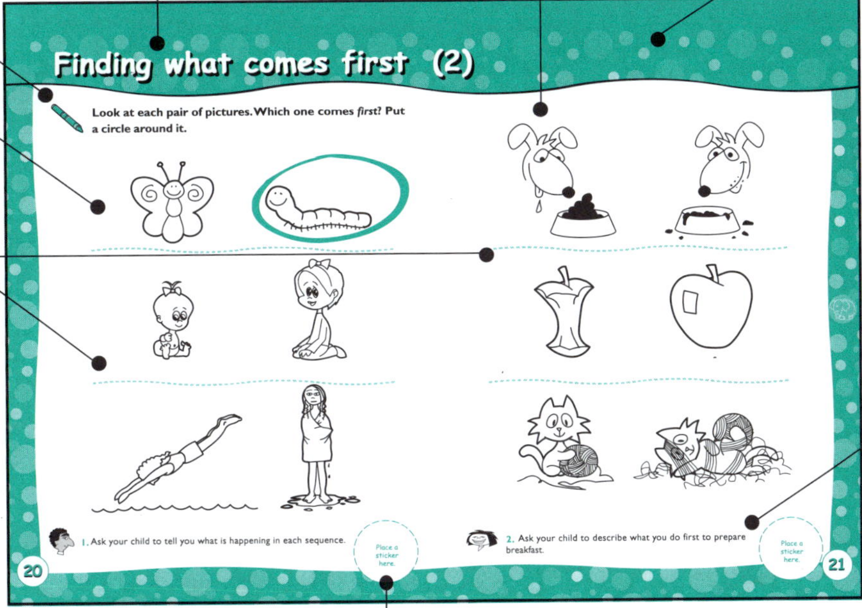

WHO'S HIDING?
In each book, a little creature appears in the border of every double page so your child can have fun trying to find it.

EXTRA ACTIVITIES
Extra activities you might want to do with your child to further reinforce the skill or simply make it more enjoyable.

Step-by-step learning

STEP ONE — **Read** out the title of the activity page to your child.

STEP TWO — **Explain** the skill and show your child the example already done. **Make sure** they understand what to do. Your child will then have at least two pages to practise that same skill.

STEP THREE — **Help** your child put a **sticker** on the bottom of each page as they complete it.

Remember to be patient, encouraging and positive with your child, even when minor mistakes are made!

How to hold a pencil

It is important that you help your child hold their crayon or pencil in the correct way, as shown here, to ensure your child develops the right technique early on.

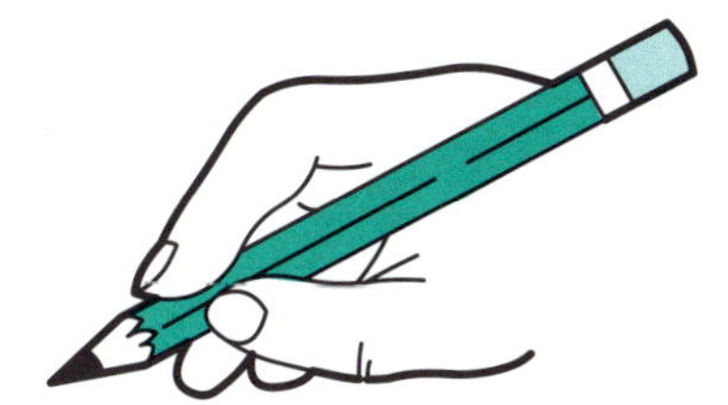

Finding one longer

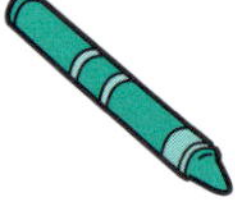

Look carefully at each pair of pictures. Which picture has something *longer* in it? Put a circle around it.

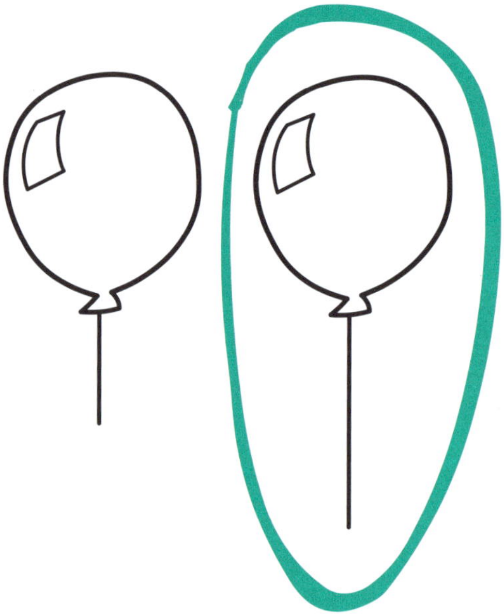

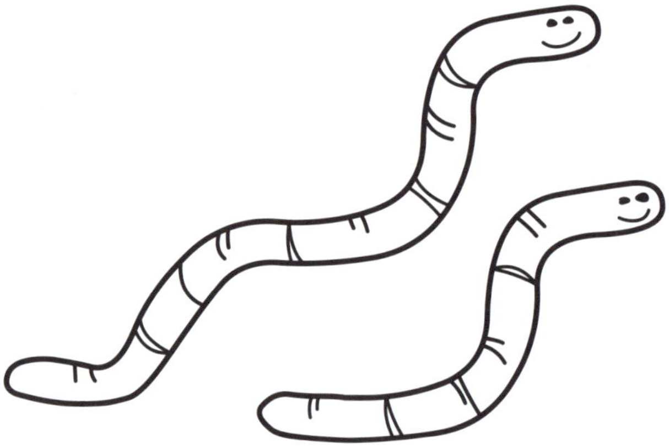

1. Ask your child to colour in the pictures which are longer.

Place a sticker here.

Great!
Great!
Terrific!
Terrific!
Great work!
Great work!
ell done!
Well done!
Fantastic!
Fantastic!
Top work!
Top work!
cellent!
Excellent!
You're a star!
You're a star!
Excellent!
Excellent!
Great!
Great!
Terrific!
Terrific!
Great work!
Great work!
ell done!
Well done!
Fantastic!
Fantastic!
Top work!
Top work!
cellent!
Excellent!
You're a star!
You're a star!
You're a star!
You're a star!
Great!
Great!
Terrific!
Terrific!
Great work!
Great work!
ell done!
Well done!
Fantastic!
Fantastic!
Top work!
Top work!

2. Cut a piece of string as long as your child's body, then help them find two things which are longer than this piece of string.

Place a sticker here.

Drawing one longer

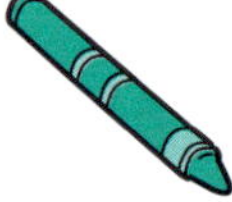

Look at each picture. Trace over the dotted lines to draw something *longer* next to each one.

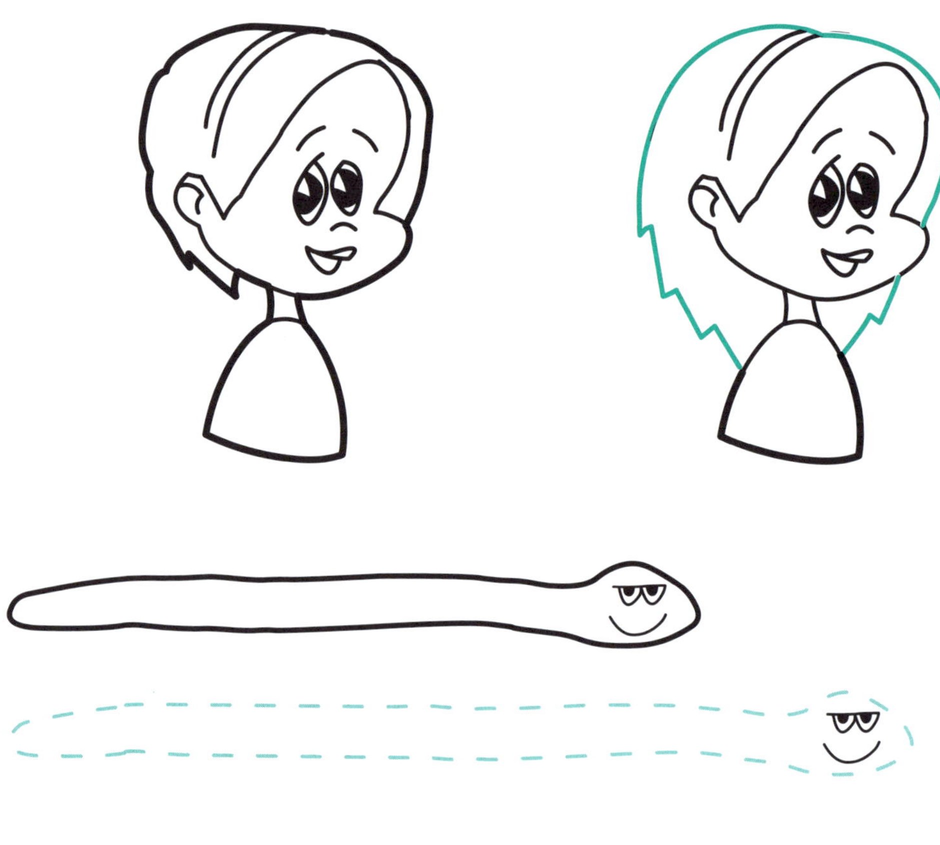

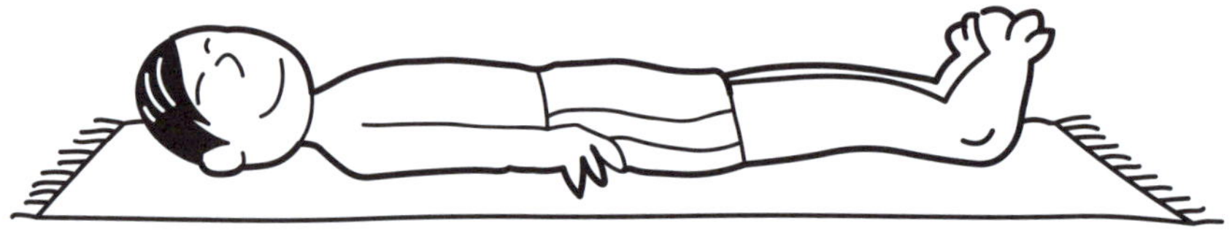

1. Ask your child to compare the hair lengths of the people around you: who has the longest hair? Who has the shortest?

Place a sticker here.

2. Have your child stand with their arms outstretched. Ask which is longer — their *reach* (outstretched arms) or their *height*? You can use a mirror to help.

Place a sticker here.

Feeling surfaces

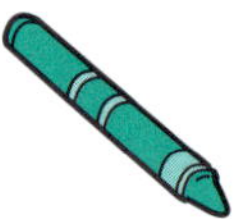

Close your eyes and rub your hand over the carpet or a rug on the floor. What does it feel like? Put your page on the carpet, then carefully run your crayon over the first box. Then do the same in the other boxes.

Carpet

Floor tile

Wall

1. Ask your child which of these surfaces feel rough, and which feel smooth.

Place a sticker here.

Table

Towel

Doormat

2. If you don't have the suggested surfaces within reach, try other ones around you with interesting textures, for example, tree trunks, any kind of wood grain, or a mesh surface.

Place a sticker here.

Comparing two areas

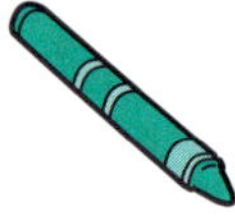

Look at each pair of pictures. Circle the one which has a *larger* area.

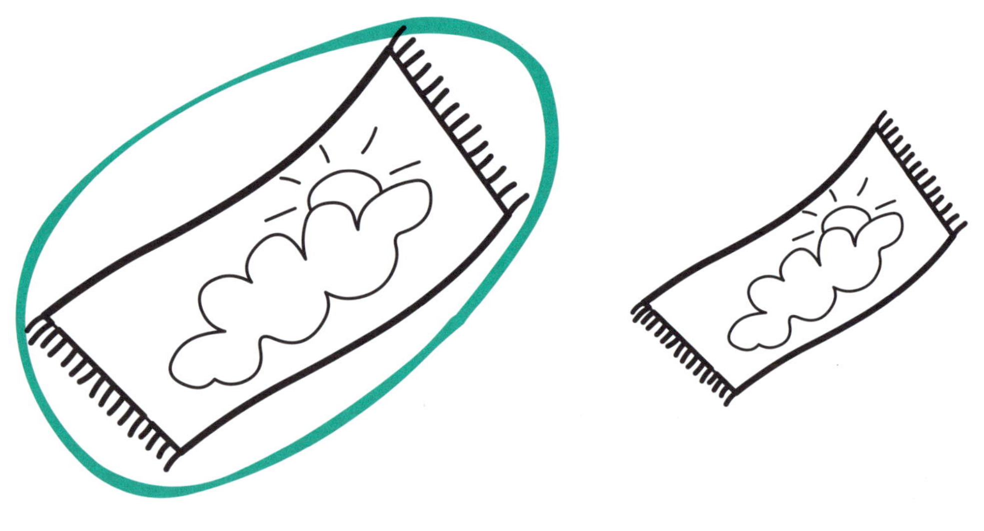

1. Your child can colour in the pictures which have a bigger area in the same colour.

Place a sticker here.

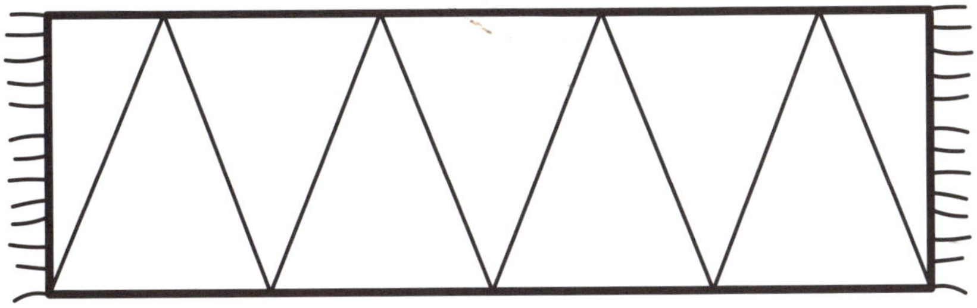

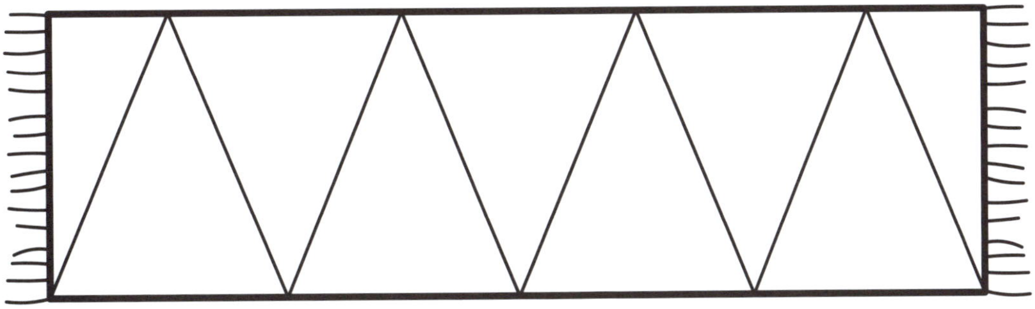

2. Compare objects around your home, for example, the top of the dinner table and the top of your child's bed.

Place a sticker here.

Finding one larger

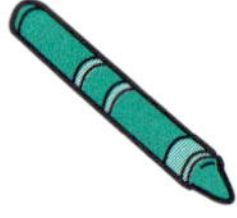

Look carefully at each pair of pictures. Circle the picture which is *larger*.

1. Your child can colour in the pictures which are larger in the same coloured crayon.

Place a sticker here.

2. Find pairs of objects around the home and ask your child which one is larger.

Place a sticker here.

Drawing one larger

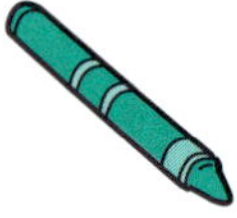

Look at each picture. Trace over the dotted lines to draw something *larger* next to each one.

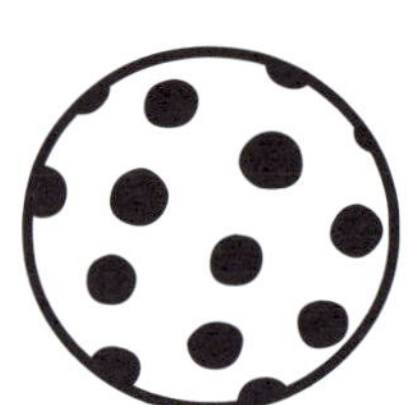

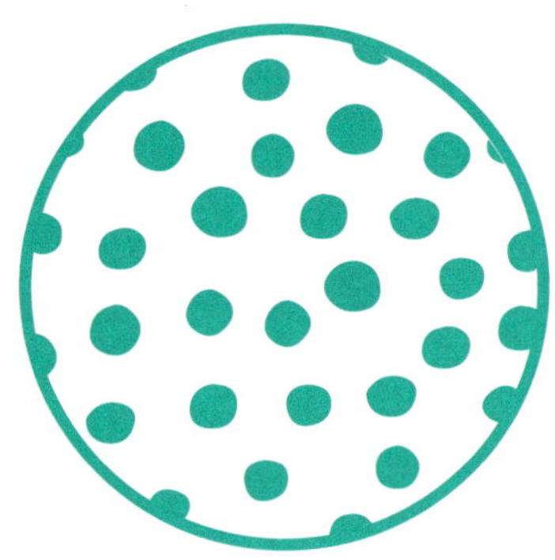

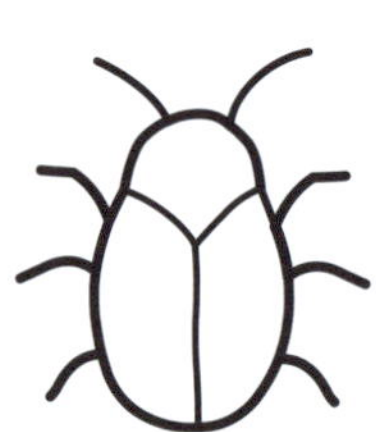

1. Ask your child to colour in all the larger things red, and all the smaller things blue.

Place a sticker here.

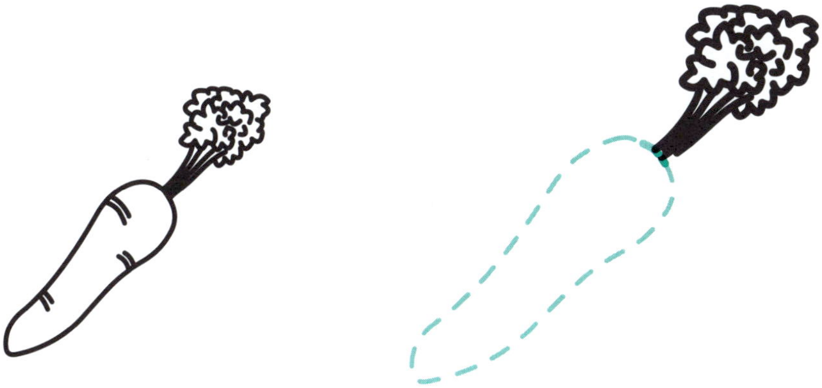

2. Ask your child to name 5 things that are larger than them.

Place a sticker here.

Finding one heavier

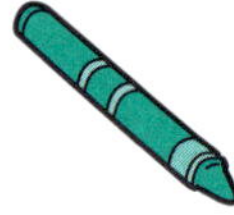

Look at each pair of pictures. Which one would feel *heavier*? Put a circle around it.

1. Select two different objects, one much heavier than the other. Have your child close their eyes and put one object in each of their hands. Ask which feels the heaviest.

Place a sticker here.

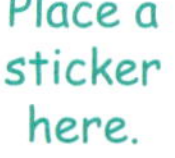

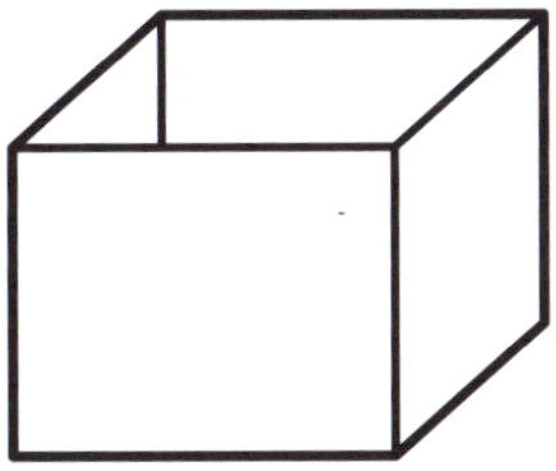

2. Take two household objects of obviously different weight and ask your child to guess which is heavier. Then let your child hold the objects and see if the guess was correct.

Place a sticker here.

Finding 'hot' things

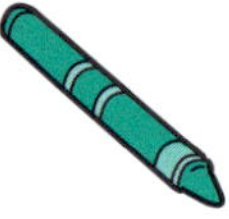

Look at each pair of pictures. Which one shows something that is *hot*? Put a circle around it.

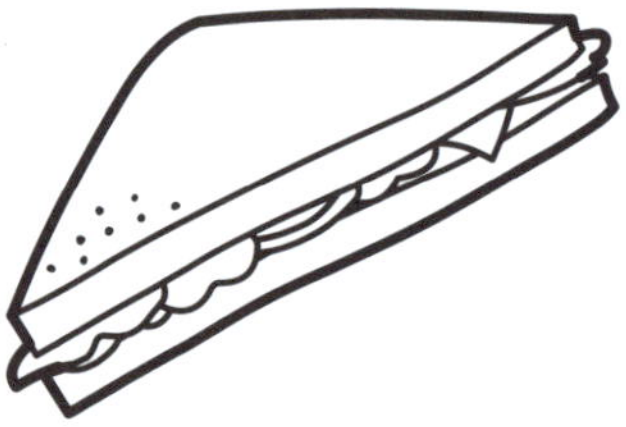

1. Ask your child to name hot things, for example, a candle flame, or your body after running.

Place a sticker here.

2. Ask your child to colour the hot things in red.

Place a sticker here.

Finding what comes first

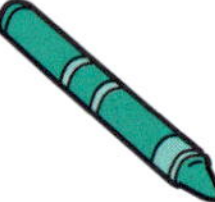

Look at each pair of pictures. Which one comes *first*? Put a circle around it.

1. Ask your child to tell you what is happening in each sequence.

Place a sticker here.

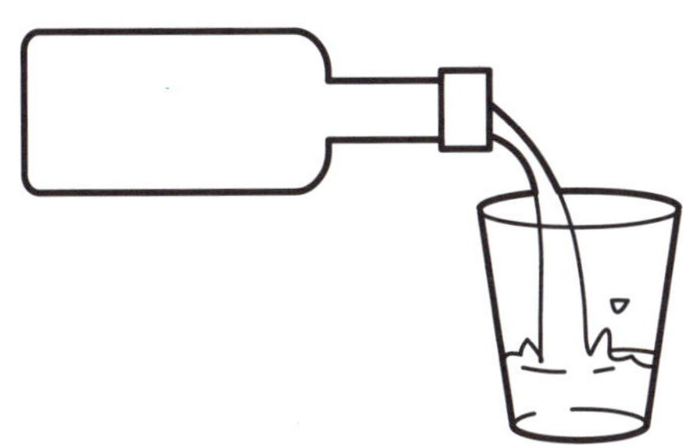

2. Ask your child what they do first thing in the morning.

Place a sticker here.

Finding what comes first

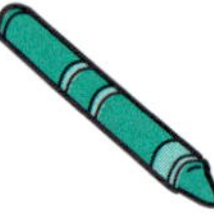

Look at each pair of pictures. Which one comes *first*? Put a circle around it.

1. Ask your child to tell you what is happening in each sequence.

Place a sticker here.

2. Ask your child to describe what you do first to prepare breakfast.

Place a sticker here.

Finding balls

What is a ball? All the pictures on the first page have balls in them. Colour all the big balls in blue. Colour the small ones in red. Then, on the next page, put a circle around any shape that is not a ball.

1. Can your child see any balls around them? Soccer balls? Rubber balls? Tennis balls?

Place a sticker here.

2. See if your child can find the balls in an assortment of real balls and other toys.

Place a sticker here.

Finding boxes

What is a box? All the pictures on the first page have boxes in them. Colour all the big boxes in orange. Colour the small ones in purple. Then, on the next page, put a circle around any shape that is not a box.

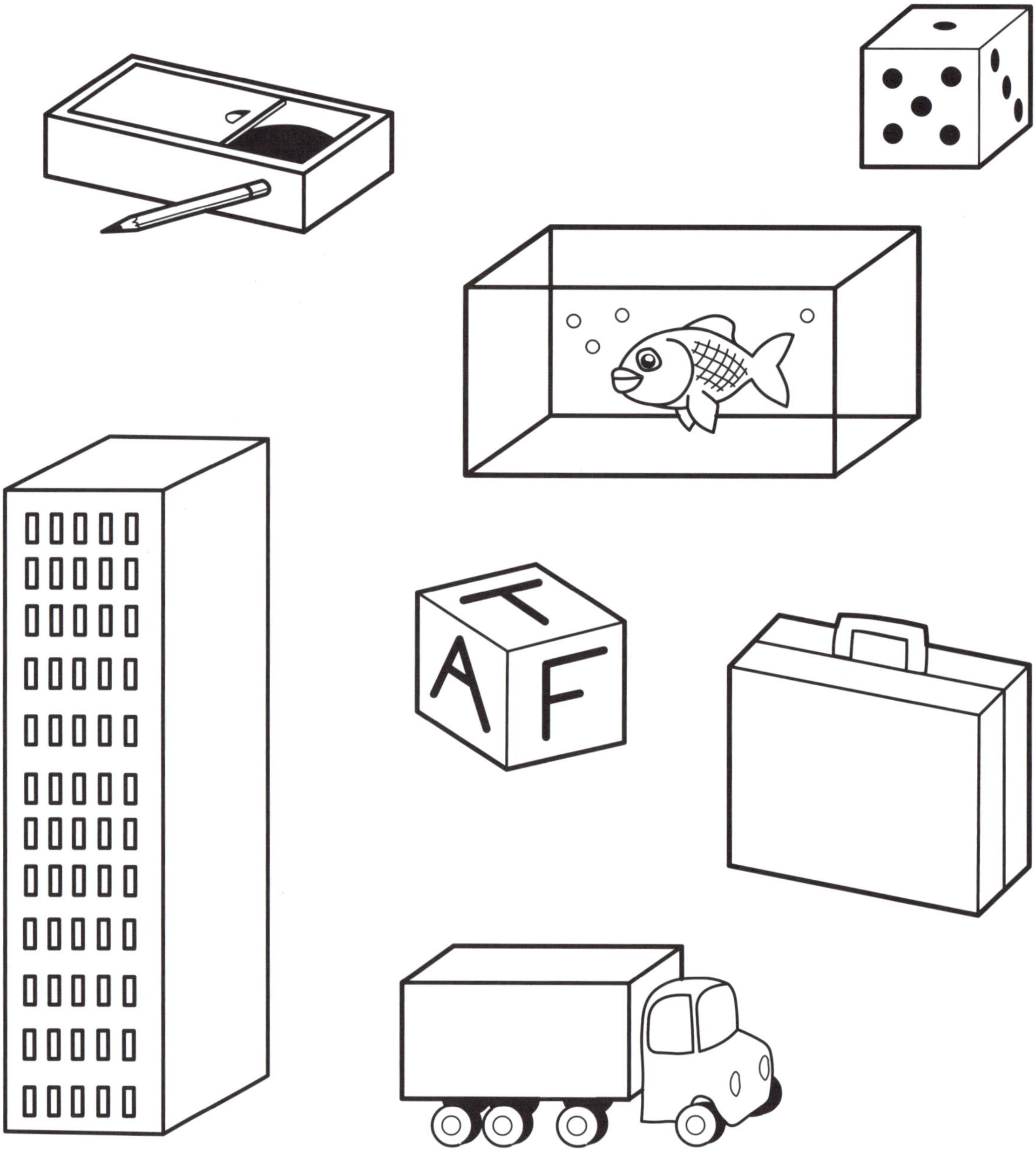

1. Can your child see any boxes around them? Packing boxes? Boxes of biscuits? A shoebox?

Place a sticker here.

2. See if your child can find the boxes in an assortment of toys.

Place a sticker here.

Sorting 3D shapes

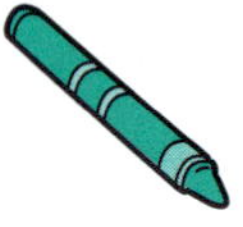

Choose two different coloured crayons, and colour in the two shapes at the top of the page. Then colour each shape in the picture to match.

1. Together you could build your own tower with things from around the home, and see if your child can find the balls and boxes in it.

Place a sticker here.

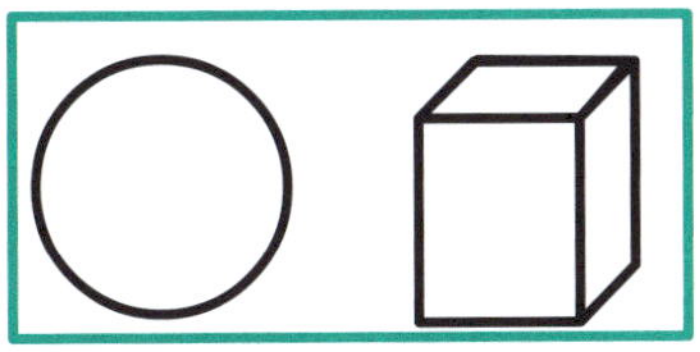

2. Explore what happens when you move a box and a ball shape. Which one rolls? Which one slides?

Place a sticker here.

Finding circles

What is a circle? All the shapes on the first dress are circles. Colour all the big circles in brown. Colour the small ones in yellow. Then, on the next page, put a circle around any shape that is not a circle.

1. Help your child draw a circle by tracing around a saucer or another round object (eg. the rim of a cup or round cake tin).

Place a sticker here.

2. Does your child have a piece of clothing with a circle on it? Can they find the circles?

Place a sticker here.

Finding squares

What is a square? All the shapes on the first shirt are squares. Colour all the big squares in red. Colour the small ones in green. Then, on the next page, put a circle around any shape that is not a square.

1. Help your child draw a square by tracing around a square object, such as a tile or a square plate.

Place a sticker here.

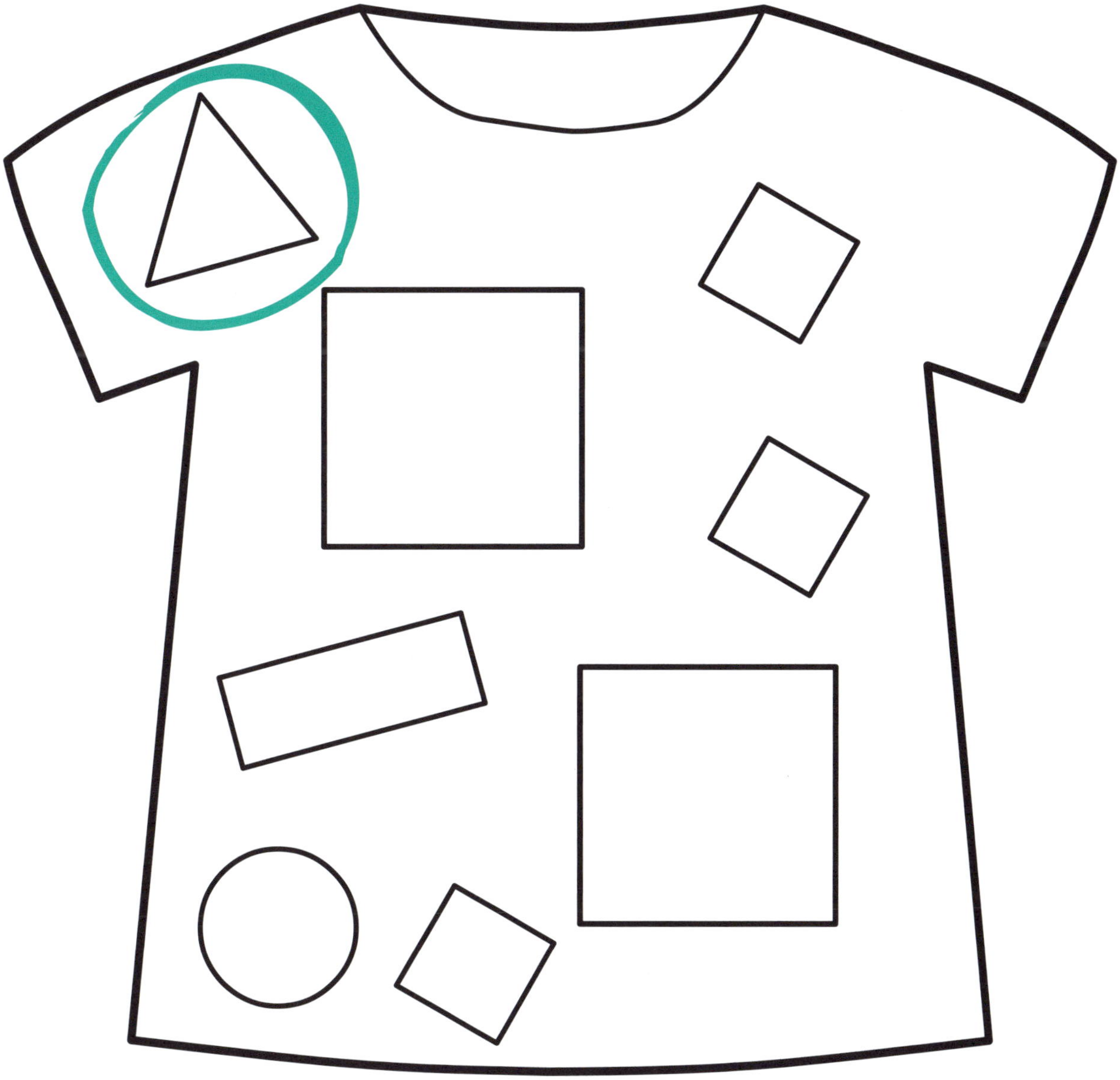

2. Does your child have a piece of clothing with a square on it? Can they find the squares?

Place a sticker here.

Finding triangles

What is a triangle? All the shapes on the first rug are triangles. Colour all the big triangles in yellow. Colour the small ones in blue. Then, on the next page, put a circle around any shape that is not a triangle.

1. Look around you for a triangle. Where can your child see one? In a piece of cheese or a wedge of pie? In toys?

Place a sticker here.

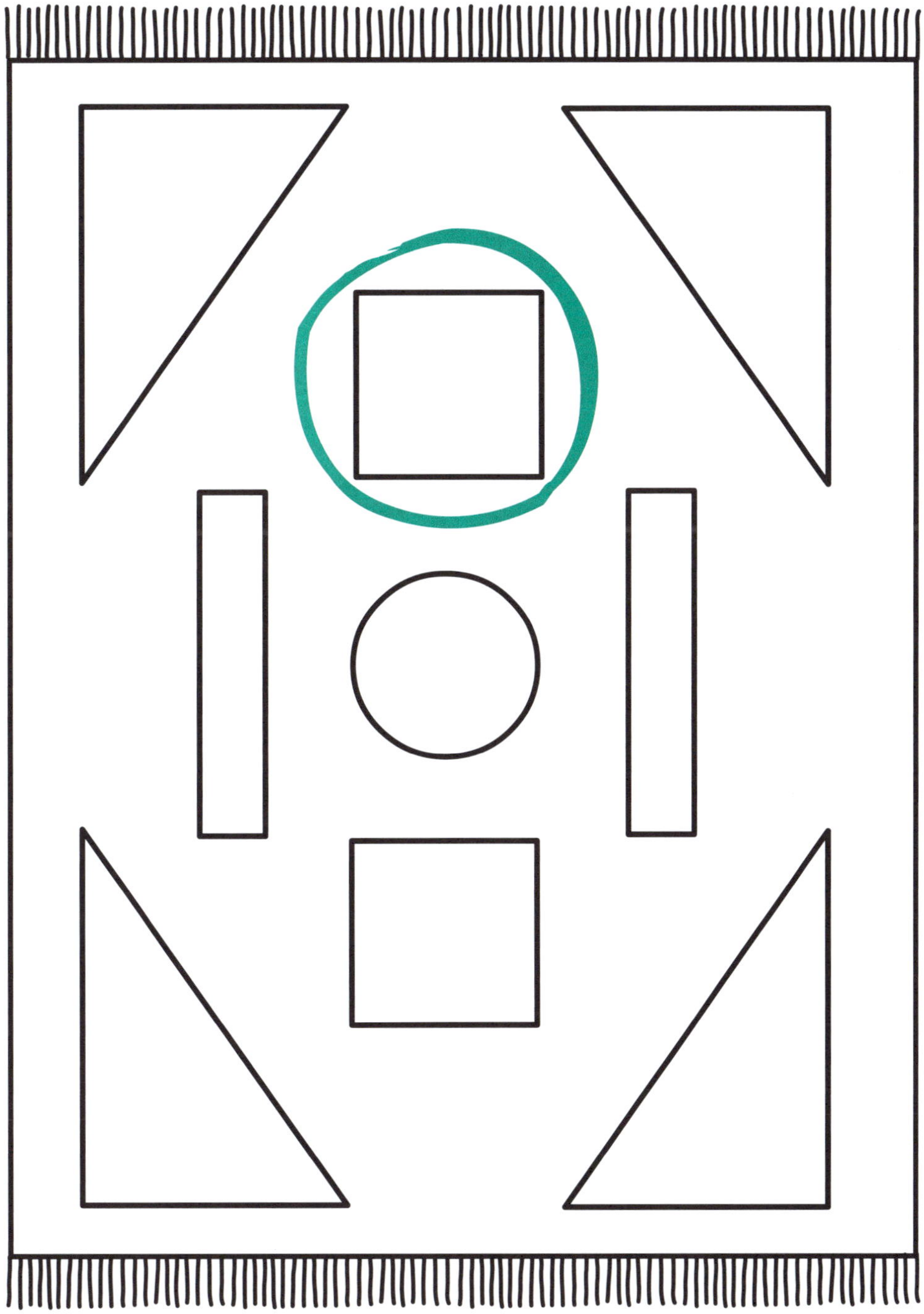

2. Help your child make their own triangle pattern by cutting out triangles from coloured paper and making patterns.

Place a sticker here.

Finding rectangles

What is a rectangle? All the shapes on the first frame are rectangles. Colour all the big rectangles in purple. Colour the small ones in pink. Then, on the next page, put a circle around any shape that is not a rectangle.

1. Look around you for a rectangle. Where can your child see one? In a piece of paper? In a picture frame on the wall?

Place a sticker here.

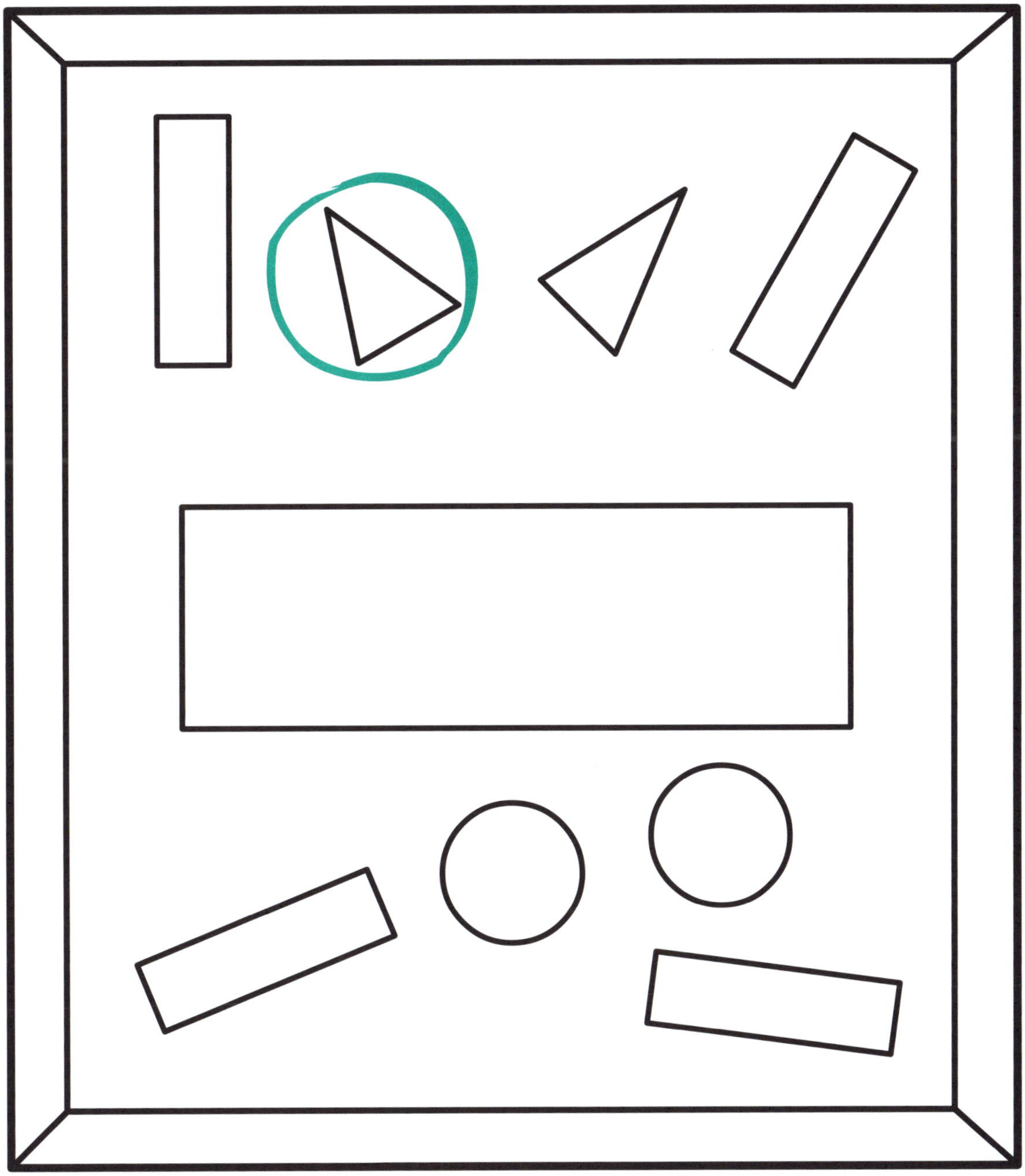

2. Go out for a walk or ride outside and look for shapes you know.

Place a sticker here.

Making patterns

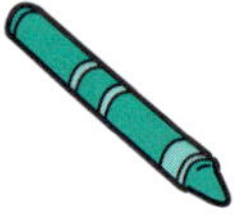

Where do you see patterns? Make your own patterns by colouring in each group in your favourite colours.

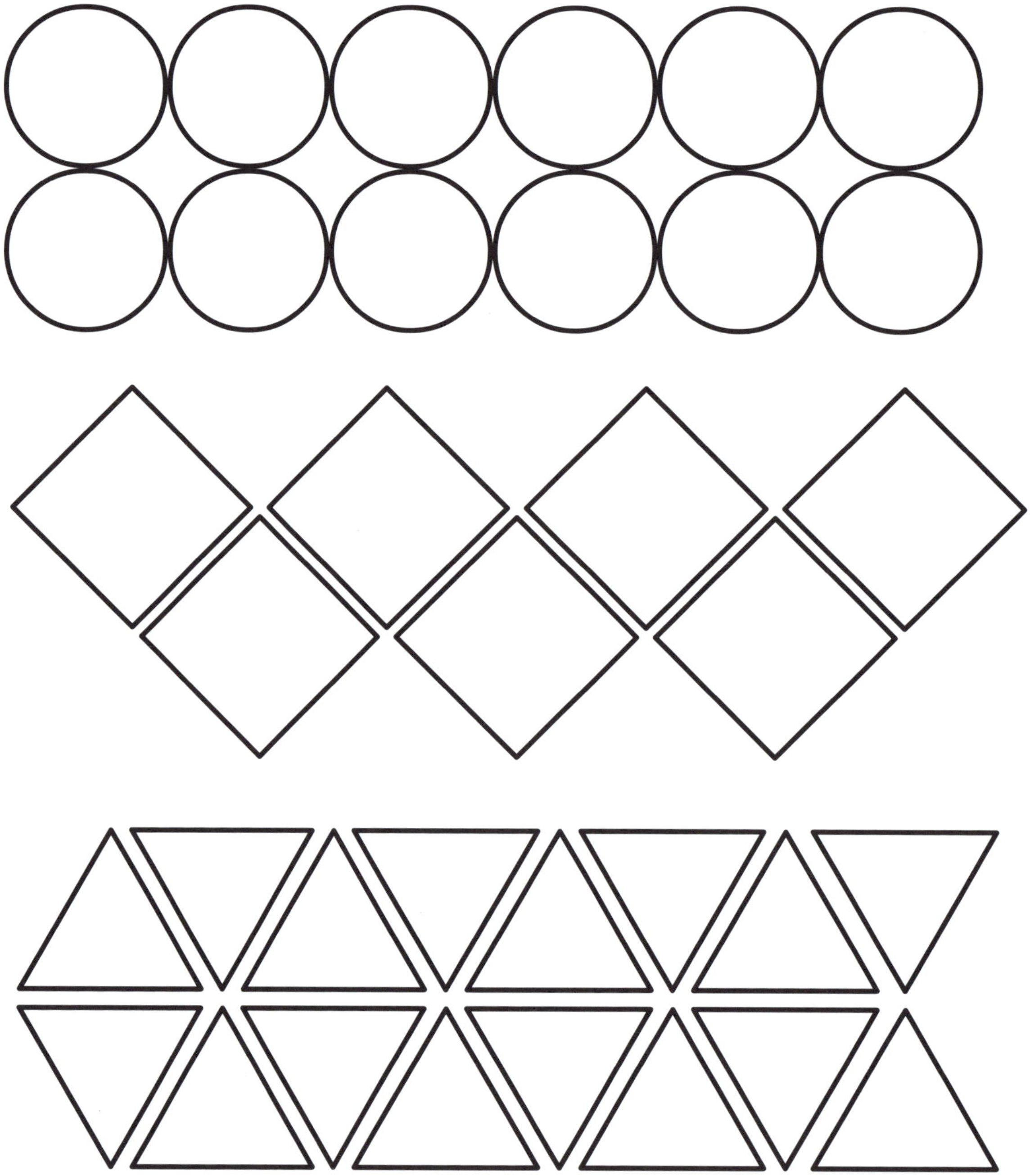

1. Where can your child see patterns around the home? In floor tiles? In their bedroom?

Place a sticker here.

2. Help your child make their own patterns by tracing around objects onto paper, then colouring them in.

Place a sticker here.

Making pictures

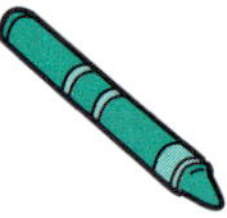

You can make pictures by putting different shapes together. Colour each picture in your favourite colours.

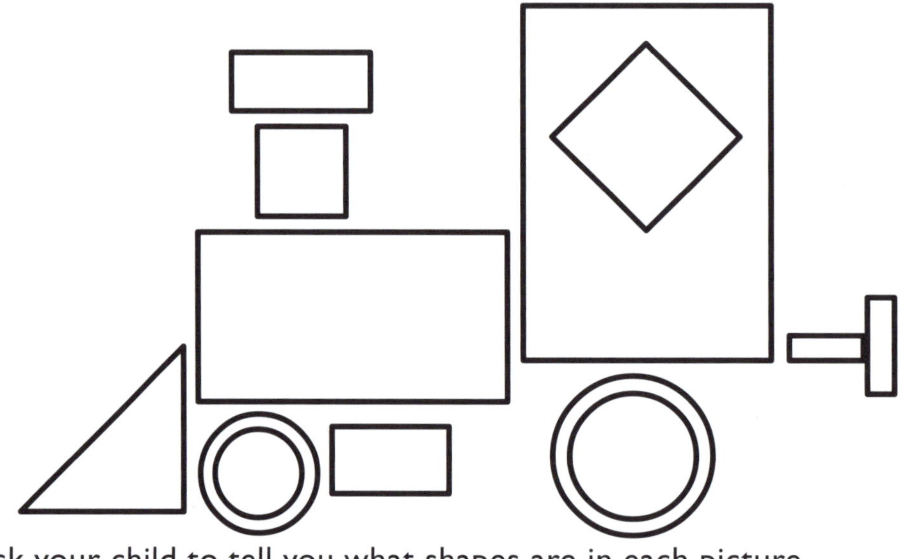

1. Ask your child to tell you what shapes are in each picture.

Place a sticker here.

2. You can help your child to make a simple picture by tracing around objects or by drawing simple shapes on paper (eg. a house with a square and a triangle).

Place a sticker here.

Drawing something 'on top of'

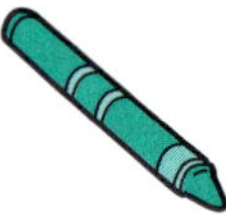

What can you get on top of? A person can be *on top* of a horse. Draw something *on top* of each animal or object.

1. Ask your child what they can get on top of — a chair? A bed?

Place a sticker here.

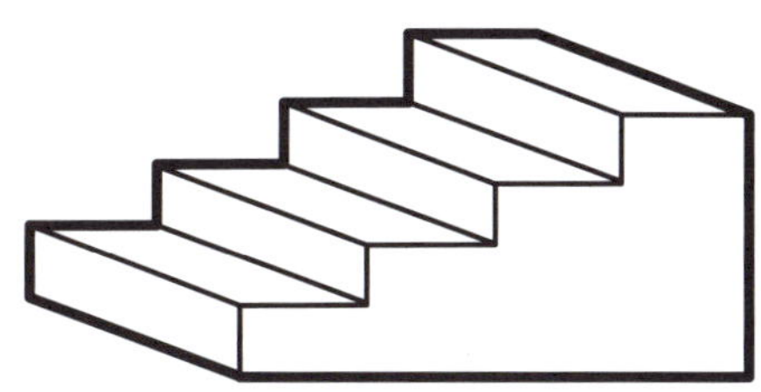

2. Ask your child to place some objects on top of others (eg. put the books on top of the table).

Place a sticker here.

Drawing something 'between'

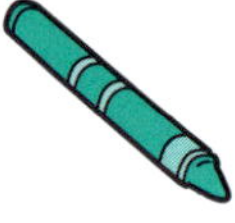

What can you stand *between*? You can stand *between* a bed and a table. Draw a shape or a picture between each two things.

1. Show your child what 'between' means by standing him or her between two things: a lamp and a chair, or between you and another object.

Place a sticker here.

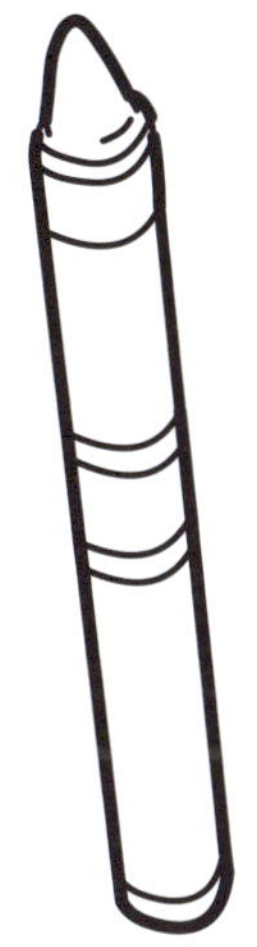

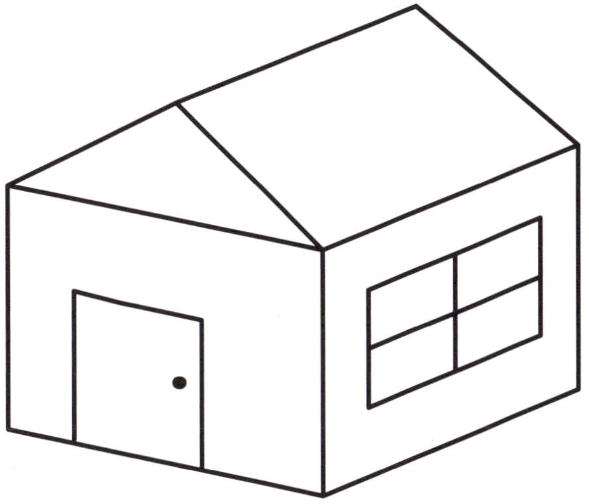

2. Ask your child to put an object between two others (eg. put a toy between a pencil and a book).

Place a sticker here.

Finding your way (1)

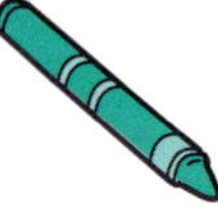

Trace the way through each maze so the rabbit gets the carrots and the horse gets the hay.

1. You might need to explain to your child how to do a maze and show them how to trace through it.

Place a sticker here.

2. Ask your child to tell you where there is a dead end.

Place a sticker here.

Finding your way (2)

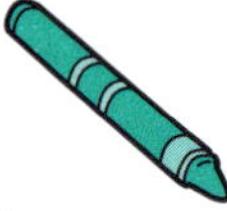

Trace the way through each maze so the monkey gets the bananas and the boy gets home.

1. Help your child by tracing different paths with your finger first.

Place a sticker here.

2. This maze may present more of a challenge for your child. Help them trace the correct path with their finger before drawing it in.

Place a sticker here.

Well done!

You have finished the book!

Place your last two stickers on the picture. You can colour in the picture, too.